SECRET

**OPERATIONAL GROUPS
FIELD MANUAL –**

STRATEGIC SERVICES
(Provisional)

Prepared under direction of
The Director of Strategic Services

SECRET

OPERATIONAL GROUPS FIELD MANUAL

– STRATEGIC SERVICES
(Provisional)

Strategic Services Field Manual No. 6

SECRET

Office of Strategic Services

Washington, D. C.

25 April 1944

This Operational Groups Field Manual — Strategic Services is made available for the information and guidance of selected personnel and will be used as the basic doctrine for Strategic Services training for the operations of these groups.

The contents of this manual should be carefully controlled and should not be allowed to come into unauthorized hands. The manual will not be taken to advance bases.

AR 380-5, 15 March 1944, pertaining to the handling of secret documents, will be complied with the handling of this manual.

William J. Donovan

Director

SECRET

TABLE OF CONTENTS

SECTION I — INTRODUCTION

1. SCOPE AND PURPOSE OF MANUAL . . . 1
2. DEFINITIONS 1
3. OPERATIONAL GROUPS 2

SECTION II — ORGANIZATION

4. ORGANIZATION IN WASHINGTON . . . 3
5. ORGANIZATION IN THE FIELD 4

SECTION III — PERSONNEL

6. ORGANIZATION FOR RECRUITMENT . . 7
7. QUALIFICATIONS OF OG PERSONNEL . . 8

SECTION IV — TRAINING

8. GENERAL PROCEDURE 10
9. TRAINING OBJECTIVES 11
10. CURRICULUM 11
11. MAINTENANCE OF MORALE 13

SECTION V — OPERATIONS

12. GENERAL 14
13. TYPES OF OG OPERATIONS 14
14. OPERATIONAL PROBLEMS 17

SECTION VI — COOPERATION OF OG WITHIN OSS AND WITH OTHER ORGANIZATIONS

15. COOPERATION WITH THE INTELLIGENCE SERVICE 19
16. COOPERATION WITH OTHER OSS OPERATIONS BRANCHES 20
17. COOPERATION WITH SIMILAR AGENCIES OF ALLIED NATIONS 20

SECTION VII — PLANNING

18. PLANNING IN WASHINGTON 21
19. PLANNING IN THEATERS OF OPERATIONS 21
20. CHECK LIST 22

SECRET

OPERATIONAL GROUPS FIELD MANUAL — STRATEGIC SERVICES

SECTION I — INTRODUCTION

1. *SCOPE AND PURPOSE OF MANUAL*

This manual sets forth the authorized functions, operational principles, methods, and organization of Operational Groups (OG's) as a part of OSS operations. Its purpose is to guide Strategic Services personnel responsible for planning, training, and operations in the proper employment of OG's.

2. *DEFINITIONS*

a. OVER-ALL PROGRAM FOR STRATEGIC SERVICES ACTIVITIES — a collection of objectives, in order of priority (importance) within a theater or area.

b. OBJECTIVE — a main or controlling goal for accomplishment within a theater or area by Strategic Services as set forth in an Over-all Program.

c. SPECIAL PROGRAM FOR STRATEGIC SERVICES ACTIVITIES — a statement setting forth the detailed missions assigned to one or more Strategic Services branches, designed to accomplish a given objective, together with a summary of the situation and the general methods of accomplishment of the assigned missions.

d. MISSION — a statement of purpose set forth in a special program for the accomplishment of a given objective.

e. OPERATIONAL PLAN — an amplification or elaboration of a special program, containing the details and means of carrying out the specified activities.

f. TASK — a detailed operation, usually planned in the field, which contributes toward the accomplishment of a mission.

g. TARGET — a place, establishment, group, or individual toward which activities or operations are directed.

SECRET

h. THE FIELD — all areas outside of the United States in which Strategic Services activities take place.

i. FIELD BASE — an OSS headquarters in the field, designated by the name of the city in which it is established, e.g., Strategic Services Field Base, Cairo.

j. ADVANCED OR SUB-BASE — an additional base established by and responsible to an OSS field base.

k. OPERATIVE — an individual employed by and responsible to the OSS and assigned under special programs to field activity.

l. AGENT — an individual recruited in the field who is employed and directed by an OSS operative or by a field or sub-base.

m. RESISTANCE GROUPS — individuals associated together in enemy-held territory to oppose the enemy by any or all means short of military operations, e.g., by sabotage, non-cooperation.

n. GUERRILLAS — an organized band of individuals in enemy-held territory, indefinite as to number, which conducts against the enemy irregular operations, including those of a military or quasi-military nature.

3. *OPERATIONAL GROUPS*

a. DEFINITION

OPERATIONAL GROUPS: a small, uniformed party of specially qualified soldiers, organized, trained, and equipped to accomplish the specific missions set forth below.

b. AUTHORITY

Among the functions assigned by Joint Chiefs of Staff directive to the Office of Strategic Services are the following, which are applicable to Operational Groups:

(1) The organization and conduct of guerrilla warfare;

(2) The use of the organization and facilities of the OSS by the theater commander in his theater or

SECRET

area in any manner and to the maximum extent desired by him.

<u>c</u>. MISSIONS OF OPERATIONAL GROUPS

The mission of Operational Groups is:

(1) To organize, train, and equip resistance groups in order to convert them into guerrillas, and to serve as the nuclei of such groups in operations against the enemy, as directed by the theater commander.

(2) In addition, under authority granted to the theater commander by the JCS Directive, Operational Groups may be used to execute independent operations against enemy targets as directed by the theater commander.

SECTION II — ORGANIZATION

4. *ORGANIZATION IN WASHINGTON*

<u>a</u>. Operational Groups are organized in Washington along strictly military lines. There is a commanding officer, responsible to the Strategic Services Operations Officer, and a staff consisting of an executive officer, an S-1 (personnel), and S-2 (intelligence and security), an S-3 (plans and training), an S-4 (supply), and a medical officer (chief surgeon and medical supply officer). There is also a training staff of variable size consisting of semi-permanent senior instructors, and junior instructors who are assigned to field duty with OG's after they have trained their successors.

<u>b</u>. OG Headquarters, Washington, has no direct command over OG's in the field, since they are under control and direction of the theater commander through the strategic services officer. The primary function of the OG organization in Washington is to service OG's in the field with trained personnel and supplies. OG Headquarters, Washington, also has the administrative responsibility of maintaining coordinated chronological record of OG activities.

SECRET

5. *ORGANIZATION IN THE FIELD*

a. THE OPERATIONAL GROUP

(1) TABLE OF ORGANIZATION

The Operational Groups, consisting of 4 officers and 30 men, is the basic unit of OG organization. An OG normally consists of 2 sections of 2 squads each. The T/O of a typical OG is as follows:

Captain (1), commanding
First Lieutenant (3), including:
  Second-in-command of the OG (1)
  Section leader (2)
Technical sergeant (2), including:
  Second-in-command of sections (2)
Staff Sergeant (6), including:
  Squad leader (4)
  Medical technician (2)
Corporal or technician fifth grade (22), including:
  Scout (16)
  Code clerk (1)
  Courier (1)
  Radio operator (4)
Aggregate (all ranks): 34

(2) TABLE OF EQUIPMENT

In addition to standard Army clothing, OG members are issued special garments appropriate to the climate and terrain in their country of operations. Each Operational Group has a special Table of Equipment (T/E), showing the arms and other articles to be carried. This T/E varies with the theater for which the OG is bound and the missions it is expected to accomplish.

(3) SS EQUIPMENT

(a) SS weapons and demolition equipment are issued to OG's through SS supply channels in the theater, as required by their missions.

(b) Communications equipment carried by OG's consists of SS radio sets which are issued through SS supply channels in the theater.

(4) MOTORIZED VEHICLES

Although motorized vehicles are not part of the organic equipment of an OG, they may be issued in the theater when required by a mission and when it is feasible to introduce and maintain such vehicles in the area of operations.

b. THE FIELD SERVICE HEADQUARTERS

(1) TABLE OF ORGANIZATION

The Field Service Headquarters (FSHQ) is the next higher echelon of command above the Operational Group. An FSHQ is roughly comparable to the Army's battalion headquarters, and the FSHQ commanding officer directs the operations of from two to five OG's. An FSHQ is normally located outside of, but in proximity to, the enemy-held territory in which several OG's are operating. However, when conditions permit, FSHQ will be established in the area of operations. The T/O consists of the following:

>Major (1), commanding
>Captain (1), medical officer
>First lieutenant (3), including:
>>Adjutant (1)
>>Communications officer (1)
>>Supply officer (1)
>
>First sergeant (1)
>Technical Sergeant (6), including:
>>Signal non-commissioned officer (3)
>>Supply non-commissioned officer (2)
>>Replacement (1)
>
>Corporal or technician, fifth grade (16), including:
>>Armorer (1)
>>Automobile mechanic (1)
>>Clerk typist (2)
>>Code clerk-courier (6)
>>Radio operator (6)
>
>Aggregate (all ranks): 28

(2) TABLE OF EQUIPMENT

In addition to standard Army clothing, FSHQ personnel are issued special garments appropriate to the climate and terrain in their country of operations. Each FSHQ has a special T/E, showing the arms and other articles to be carried. This T/E varies with the theater in which the FSHQ is to operate and the missions it is expected to execute.

(3) SS Equipment

(a) SS weapons and demolitions equipment are issued to an FSHQ as required, through SS supply channels in the theater.

(b) Communications equipment for an FSHQ consists of SS radio sets which are issued through SS supply channels in the theater.

(4) Motorized Vehicles

Motorized vehicles are part of the organic equipment of an FSHQ and are issued through SS and military supply channels in the theater, provided it is feasible to introduce and maintain such vehicles in the area of FSHQ operations.

c. Area Headquarters (Headquarters at OSS Field Base)

(1) Table of Organization

An Area Headquarters (AHQ), or Headquarters at OSS field base, is the next higher echelon of command above the FSHQ. It operates under direction and control of the SS officer at the OSS field base. The normal T/O of an AHQ is as follows:

Lieutenant Colonel (1), OG commanding officer
Captain (1), executive officer
First lieutenant (1), operations officer
First sergeant (1)
Technical sergeant (1), signal non-commissioned officer
Corporal or technician, fifth grade (3), including:
Clerk-typist (1)
Code clerk-courier (1)

SECRET

Motorcyclist (1)
Aggregate (all ranks): 8

(2) TABLE OF EQUIPMENT

In addition to the standard Army clothing issued to personnel of the AHQ, each AHQ has a special T/E, showing the arms and other articles to be carried. This T/E is variable, depending on the theater of operations.

(3) SS EQUIPMENT

(a) Stockpiles of SS weapons and demolitions equipment are normally set up at an AHQ to supply Field Service Headquarters and OG's in areas of operations.

(b) Since the AHQ is located at an OSS field base, communications to and from AHQ are generally handled by the field base message center.

(4) MOTORIZED VEHICLES

Motorcycles, trucks, and trailers needed for operations at AHQ are supplied through SS and military supply channels in the theater.

SECTION III — PERSONNEL

6. *ORGANIZATION FOR RECRUITMENT*

a. Members of OG's procured in the United States are officers or enlisted men who have been inducted into the Army through regular channels. Under War Department approval, and within War Department allotment of grades and ratings, selection is made of such personnel by trained interviewers of the Personnel Procurement Branch (PPB), OSS, according to specifications submitted by Headquarters, Operational Groups, Washington. PPB interviewers examine the civil and military records of likely candidates and hold personal interviews. Candidates who are acceptable are ordered to an SS area to begin training, pending security clearance. This procedure in no way violates security, as the

SECRET

training initially given is an extension of Army training. No specialized strategic services instruction is given until the security check has been completed.

b. It will sometimes be necessary to procure OG personnel directly in the theater where they will operate. This procedure is applicable when persons cannot be found in the United States who are qualified in a particular language, knowledge of a certain locality, and other essentials. When an OG must be staffed in the theater, the work of procurement will usually be done by a cadre from the U.S. consisting normally of 2 officers and 5 men, with the following T/O: 1 captain, 1 first lieutenant, 1 first sergeant, 1 staff sergeant, 1 sergeant, and 2 radio operator technicians, fourth or fifth grade (specification serial No. 777). This cadre will attempt to recruit and train in the theater sufficient personnel to comprise standard OG's of 4 officers and 30 men each. However, the T/O may be reduced in strength for OG's recruited in the field depending on the availability of qualified personnel. Civilians recruited for OG's in the field will be enlisted or commissioned in the Army of the U.S. and will wear its uniform. The procurement of all personnel for OG's within theaters must be within the limitations of authorized grades and ratings.

7. *QUALIFICATIONS OF OG PERSONNEL*

The following considerations will govern selection of personnel for Operational Groups:

a. WILLINGNESS TO PERFORM HAZARDOUS DUTY

Because of the nature of their assignments, all members of OG's must be willing to undertake unusual and dangerous risks. Candidates must be adequately informed of the hazards they may expect, and must be accepted only on a volunteer basis.

b. LANGUA ABILITY

the rmally preferable that the candidate speak language as a native tongue, or with great fluency. In some cases, however, e.g., radio operators, language facility must be sacrificed for other valuable qualifications.

c. FAMILIARITY WITH COUNTRY OF OPERATIONS

Since OG's may have to enter territory without benefit of a friendly local reception committee, previous acquaintance with the country of operations is highly desirable, especially if such acquaintance is of recent date. OG personnel with friends or relatives who might provide concealment and guidance are especially valuable.

d. SKILLS

As many men as possible in each OG should be qualified in certain specialized fields. Previous training on radio, demolitions, weapons, scouting, or fieldcraft is a particularly desirable qualification in a candidate.

e. PHYSICAL CONDITION

The rigorous character of their work demands that OG personnel satisfy the same physical requirements as men accepted for parachute training in the Army.

f. POLITICAL SYMPATHIES

Persons charged with procurement of OG personnel must use great care in the case of individuals who are sympathetic to particular political movements or factions within the country of their origin. The readiness and ability of such individuals to get along harmoniously with the movement or faction in the area of operations must be carefully determined in advance. In certain areas, however, where disputes are bitter, and the areas of rivals not delineated, it is more desirable to staff an OG with American citizens whose language ability is somewhat imperfect rather than with ex-natives of the area who have pronounced political attachments.

g. CHARACTER TRAITS

While the risks involved tend to make OG work appeal to young men, the success of OG assignments

SECRET

is not the result of daring and bravado alone. Accordingly, candidates will be selected whose past records, civilian and military, give evidence of stability and good judgment.

h. ARMY TRAINING

Except for certain specially qualified persons recruited in the field (see paragraph 6.b.) candidates must have completed basic training before being accepted for OG work. Candidates who have also had combat training are preferable.

SECTION IV — TRAINING

8. *GENERAL PROCEDURE*

OG training is an intensive course of specialized instruction in the weapons, techniques, and methods of operation appropriate for a small, self-sufficient band of men who may be required to live and fight in the manner of guerrillas. OG training comes under the general supervision of the Schools and Training Branch, but the actual instruction is given by OG personnel, based on schedules drawn up by the OG training officer. An OG is assembled prior to the start of training according to the common foreign language of its members; thereafter, the group trains, lives and operates as a unit. The officers who will lead an OG in the field assists in training its personnel. The training period in the U.S. is normally three weeks. One additional week is allowed for the clearance of administrative details. The group is then ready for embarkation to the theater of operations. An OG is rarely used immediately upon its arrival overseas. The normal time delay involved is utilized for further training, as dictated by the particular mission to be performed. This training will emphasize tactical problems and may include parachute jumping or amphibious operations if either of these means of entry is to be used. Overseas training is usually conducted by OG officers.

SECRET

9. *TRAINING OBJECTIVES*

The objectives of OG training are as follows:

a. To train specially qualified bi-lingual officers and enlisted men in the techniques and skills required to execute their prescribed missions in enemy or enemy-occupied territory.

b. To weld this personnel into an efficient, mobile, self-sufficient unit capable of:

(1) organizing and training local resistance groups with a view to converting them into guerrillas;

(2) supplying such guerrillas withs arms, ammunition, demolition, communication equipment, food, medical supplies, and money;

(3) serving as nuclei in planning and execution by native elements of attacks against enemy forces or installations, as directed by the theater commander;

(4) executing independent operations, usually of a "hit-and-run" character, against enemy targets as directed by the theater commander.

c. To develop in each member of an Operational Group the physical strength, individual initiative, and ability to improvise, which his missions will demand.

10. *CURRICULUM*

a. Members of Operational Groups should receive adequate training in the following subjects:

(1) Map study, including map sketching map-and-compass problems, direction-finding by field expedients, study of aerial photos.

(2) Scouting and patrolling, including instruction and practice in use of physical cover, reconnaissance, signalling, infiltration.

(3) Close combat (armed and unarmed), including knife-fighting.

SECRET

(4) Physical conditioning, including swimming, toughening exercises, and obstacle course runs.

(5) Fieldcraft, including camouflage, living off the land, preparation of shelter and food.

(6) Hygiene and camp sanitation.

(7) Tactics, including basic maneuvers and tactical principles, discussion and practice in small-group operations and methods of guerrilla warfare, day and night problems, planning and execution of airborne raids, street and village fighting, jungle fighting (when applicable).

(8) Demolitions, including explosives, incendiaries, booby traps, field expedients, delayed action charges, multiple charges, charges for special purposes.

(9) Weapons, including function, stripping, cleaning, and firing of .30 cal. M1 rifle, cal. .30 carbine, cal. .30 machine gun, cal. .50 machine gun, Browning automatic rifle, cal. .45 pistol, Sten gun, cal. .45 submachine gun, grenade launcher, 2.36-inch anti-tank rocket launcher (bazooka), Marlin submachine gun, 60 mm. mortar, 81 mm. mortar, hand grenades. Also the function and firing of enemy weapons with which group may come into contact.

(10) Principles and practice of first aid, especially under combat conditions.

(11) Enemy motor transportation, including operation and repair of enemy motorcycles, trucks, automobiles, half-tracks, and other vehicles with which group may come into contact.

(12) Enemy organization, including lectures on enemy military and political structure, uniforms, insignia, procedure in interrogating prisoners, methods of espionage and counter-espionage.

(13) Methods of organizing and training civilians in the techniques of guerrilla warfare; indoctrination as to correct general attitude and behavior toward the civilians.

(14) Identification of enemy and Allied planes, tanks, and other vehicles.

(15) Care of clothing and equipment.

(16) Security, including precautions to be observed in U.S., in the theater, and in area of operations.

(17) Problems of supply, including the procedure of procuring supplies from OSS stocks, methods of packaging, and the details regarding the introduction and receipt of cargo into the zone of operations.

b. The basic training of OG's preparation will be supplemented in the theater immediately prior to operations by a detailed briefing on topography, battle order, friendly and hostile groups that may be encountered, and other matters pertinent to the operation to be performed.

c. In addition to the training outlined in paragraph a. above, radio operators for each OG should receive intensive practice in code, operational procedure, and repair of their equipment.

11. *MAINTENANCE OF MORALE*

In view of the extreme hazards of OG operations, maintenance of morale assumes a special importance. Every effort should be made throughout the training period to keep the aggressive spirit and confidence of OG personnel at a high level. The men should be kept steadily occupied, either with training tasks or with organized group recreation. Following the completion of their training, OG's will be shipped to their theater of operations as expeditiously as possible, to avoid the staleness and dissatisfaction which inevitably result from idleness or a monotonous repetition of training. All means available will be used to foster intimate friendship, mutual confidence, and teamplay among members of the group, and a strong feeling of trust between officers and men.

SECRET

SECTION V — OPERATIONS

12. *GENERAL*

<u>a</u>. OG's operate only in enemy or enemy-occupied territory. Their primary function is in connection with guerrillas. They have no operational function in neutral territory.

<u>b</u>. The following operational distinctions exist between the OG and SO Branches:

(1) While SO operating personnel may or may not be members of the armed forces, may or may not be in uniform, and operate as individuals or in small groups, OG personnel are always members of the armed forces, always operate in uniform, and conduct operations as a unit. When any individual OG personnel is selected to perform SO tasks and function, he will operate under cover and will become part of SO personnel.

(2) OG's, being military organizations, operate in accordance with military principles and on occasion will deliberately engage hostile armed forces. On the other hand, SO personnel in enemy-held territory operate under cover, except in unusual circumstances, and attempt to avoid all contact with enemy forces.

(3) Both OG and SO personnel deal with resistance groups. SO carries on a strictly covert relationship with such groups and organizes them for such tasks as attritional sabotage. OG's on the other hand, train, organize, and equip resistance groups to operate as guerrillas against enemy forces.

13. *TYPES OF OG OPERATIONS*

<u>a</u>. As set forth in paragraph 3-<u>c</u>, OG's have two broad missions. These missions determine the pattern of their operations.

(1) The primary mission of OG's is to organize, train, and equip resistance groups in order to convert

SECRET

them into guerrillas, and to serve as the nuclei of such groups in operations against the enemy as directed by the theater commander.

(a) *Organizing*

Normally before OG's enter a territory contact must have been established with resistance elements, and their potentialities and needs for supplies and equipment ascertained. This can be accomplished by use of OSS clandestine agents, primarily SO, or by representatives abroad of resistance elements who are brought out for this purpose. Such resistance elements range from small, loosely organized and poorly equipped bands of individuals to large quasi-military organizations with insufficient equipment. When organization is inadequate, the main function of OG's is to weld the individuals into a guerrilla unit that can contribute to the support of military operations. Organizing such guerrilla units may involve selecting leaders, assigning individuals or units to various areas of operation, constituting demolition or sabotage teams as the situation may require, providing for communications and courier services. While providing guidance and over-all direction is an OG responsibility, the actual leadership will usually be entrusted to local individuals. Where guerrilla activity is already well developed, the OG's work of organizing consists primarily of coordinating the operations of guerrilla bands with allied military plans. In certain areas, OG's may encounter guerrillas whose effectiveness is reduced by partisan differences. Although OG's will avoid local political controversy and will emphasize their essentially military role, they may, by their ability to furnish supplies, be effective in achieving a measure of coordinated effort among estranged groups.

(b) *Training*

The work of OG's will be mainly with civilians who are largely ignorant of military discipline, tactics, and weapons. Briefly stated, the training objective of OG's is to transform these resisters into efficient guerrillas. Within the limitations of local conditions, OG's must find ways to instruct and give practice to the patriots in such subjects as the use of weapons, close combat, scouting and patrolling, demolition, radio operation, first aid, sabotage, and physical conditioning. For obvious reasons this training should, if possible, be conducted in areas unoccupied by enemy troops, such as isolated mountain or forest regions. One of the most important OG training tasks will be indoctrinating civilians in the necessity of avoiding premature action and preserving their numbers for coordinated use at the proper time.

(c) *Equipping*

Need for additional equipment will often arise after arrival of OG's in the zone of operations. OG officers will transmit requisitions or requests for requirements either to their Field Service Headquarters, if established, or the OSS field base.

(d) *Serving as Operational Nuclei for Guerrilla Warfare*

In theaters where active military operations are being conducted, the plans covering guerrilla operations, including supply, must be approved by the theater commander. In areas where military operations are not being conducted, the nature and timing of guerrilla operations conducted by native groups under OG direction will be coordinated insofar as possible with the desires of the theater commander. In some areas it may be desirable to attack industrial or other targets at the earliest possible moment; in other regions, the theater commander may consider it essential for the groups to remain inactive until they can be employed in support of Allied military operations. In either case, the authorized function of OG's is to serve as the core of a larger group composed pre-

SECRET

dominantly of members of the local population. As indicated in paragraph 5, OG's are sub-divided into sections and squads. These smaller units will attempt to insure proper leadership and guidance for the native guerrillas whom they have trained. Typical operations by these groups might include tasks such as attacks on and demolition of a powerhouse or oil dump as well as the marking and holding of landing beaches and cutting of enemy communications.

(2) The secondary mission of OG's is to execute operations, usually of a "hit-and-run" character, against enemy targets as directed by the theater commander. It will be seen that this mission takes in a broad range of activity. Thus, OG's might conceivably be used by the theater commander to: attack an enemy headquarters; harass an enemy withdrawal; destroy enemy stores; blow up a factory; demolish a radar installation—or any one of a number of similar tasks. It is characteristic of OG operations under this category that they may or may not be closely tied in to large-scale military operations.

14. *OPERATIONAL PROBLEMS*

a. Contact with Resistance Groups

OG's assigned to organize and train resistance elements into guerrillas usually will enter the area of operations only after preliminary contact thru clandestine agents with such groups has been established and arrangements made for reception of the OG's. This contact may be made by SO or SI operatives or agents.

b. Entry into Area of Operations

The manner of entry will be determined by the terrain of the area of operations, the tightness of enemy surveillance, and the transportation available. Entry may be made by parachute, by small boat or submarine, or by infiltration of an enemy area on foot. OG's will be given special training in the theater, appropriate to the

SECRET

means of entry chosen. OG's will normally be received and guided at the point and time of entry by sympathizers with whom contact has previously been established (paragraph a above).

c. COMMUNICATIONS

As soon after entry as is feasible, OG's will make radio contact with FSHQ, if established, or with the base, according to an arranged schedule for periodic future contact. Communications will be maintained by FSHQ with all OG's functioning in the area of operations which it controls, as well as with Area Headquarters at the field base. Messages to and from Area Headquarters are handled by the field base message center. OG's will not normally attempt to communicate with any higher echelon than FSHQ. When an OG is divided into squads which operate in separate parts of the same area, contact may have to be maintained with the commanding officer of the group by radio. However, because of the risks of location by the enemy, radio traffic should be kept to a minimum. Elements of an OG may find it possible to keep in touch with each other more securely by establishing a courier service, utilizing local civilians, rather than by using radio. OG's in enemy-held territory will normally operate on foot, although in some isolated areas enemy surveillance may be so light as to permit a limited use of horses or even local motorized vehicles.

d. SUPPLY

OG's usually carry into an area of operations only such equipment as they need for their own use. OG's will survey the local status of supply, and, basing estimates upon needs previously reported and consequent preliminary plans for supply of the resistance forces, will report any additional immediate requirements by priorities. They should also report on whether the previously agreed place and means for introduction of supplies is feasible and should furnish necessary modifications.

e. CONCEALMENT

Since OG personnel operate in uniform they must

rely on concealment and secrecy to safeguard their operations. Concealment is of particular importance to OG's because they are small in number and can be severely weakened by the loss of even a few men. Prior to their entry, OG's should be issued camouflage clothing appropriate to the season and terrain. OG's will be obliged in most cases to avoid cities and towns where the enemy or his agents may be encountered. Semipermanent concealment in mountainous or forested areas may be available, and native sympathizers will be induced to provide hiding-places in their homes and barns when this is feasible. In some areas enemy controls may be so rigid as to compel OG's to keep on the move, changing bivouac sites frequently.

f. SECURITY

The enemy has established efficient espionage and counter-espionage organizations in all the occupied countries. These networks, coupled with the enemy-controlled local police and local informers, will frequently be more dangerous to the security of OG's than will the enemy's regular troops. Before OG's enter an area all possible investigations will be made as to the security of the resistance groups with whom OG's are working, but OG's must be alert to the danger of possible penetration by enemy agents. OG's should have contact only with those individuals whom resistance group leaders can personally vouch for as loyal. So far as possible, the location and operations of OG personnel should be kept secret from the families of resistance group members who are being trained and organized by OG's.

SECTION VI — COOPERATION OF OG WITHIN OSS AND WITH OTHER ORGANIZATIONS

15. *COOPERATION WITH THE INTELLIGENCE SERVICE*

a. The planning and execution of OG missions are based upon reliable intelligence, furnished primarily by the SI and X-2 Branches and the Research and Analysis Branch (R&A). Liaison between OG and these branches is maintained in Washington and in the theaters; it is most important in the theaters because all operational planning for OG's is done there. R&A provides basic intelligence with respect to topography, industrial targets, the structure of enemy military and political organization, and the attitudes of the people in the area of OG operations. For briefing purposes, SI furnishes up-to-the-minute intelligence concerning locations of enemy units and installations in the area of operations, the strength, location, and personnel of guerrilla and resistance groups that will be encountered, and such other data as is pertinent to the mission at hand. X-2 supplements this with intelligence regarding enemy espionage agents and networks which may jeopardize OG operations.

b. Although procurement of intelligence is not normally an OG task, OG's functioning behind enemy lines will frequently obtain information by reconnaissance and from the local population which will be relayed through channels to the appropriate OSS and military intelligence organizations.

16. *COOPERATION WITH OTHER OSS OPERATIONS BRANCHES*

a. OG must work in closest collaboration with SO. The integration of OG operations with this branch is achieved in Washington through the strategic services Operations officer, and in the field by the strategic services officer for each theater.

b. OG operations must be largely dependent upon SO operatives and agents who develop preliminary contacts with and make preliminary investigations of underground resistance groups prior to the entry of OG's into an area of operations.

17. *COOPERATION WITH SIMILAR AGENCIES OF ALLIED NATIONS*

Cooperation as to any joint activities with other Allied organizations conducting irregular warfare will be arranged through the strategic services officer.

SECTION VII — PLANNING

18. *PLANNING IN WASHINGTON*

a. Special Programs covering OG activities in a Theater of Operations are incorporated into OSS Over-all Programs. In the Over-all Program for a given theater, the objectives for all the OG branches concerned are set forth in order of importance. The Special OG Programs state the missions to be performed by OG to attain the objectives listed in the Over-all Program, present a brief summary of the situation bearing on the missions in question, and prescribe in a general way the plan to be followed. These Special OG Programs are drawn up jointly by the Strategic Services Planning Staff and the OG Branch, and are presented to the Strategic Services Planning Group for approval. Upon approval by the Planning Group, the Programs are submitted to the Director, OSS, for his consideration and approval before being transmitted to the theater or senior American commander in the field through the strategic services officer.

b. Upon approval of theater commanders, OG Programs establish priorities for OG operations in the field. In conformity with these programs, OG prepares detailed operational plans.

c. When plans covering OG activities in the field are made which are not in furtherance of missions set forth in Special Programs, such plans are reported to OSS, Washington, for consideration and incorporation into an appropriate program, consistent with security control.

19. *PLANNING IN THEATERS OF OPERATIONS*

Operational planning for OG's is performed in the field, in the implementation of missions of approved special programs covering OG activities. Such planning should cover the details listed in paragraph 14. The nature of OG operations makes teamwork essential and requires that planning be executed in the most minute detail possible.

SECRET

20. *CHECK LIST*

In Appendix "A" there are summarized in the form of a check list a number of the more important points that have been presented in this manual. This check list may serve as a brief list of reminders to OG personnel to assist them in the course of their work.

APPENDIX "A"

TO

OPERATIONAL GROUPS FIELD MANUAL —

STRATEGIC SERVICES

CHECK LIST FOR OG OPERATIONS

This check list is designed to assist the OG Branch, Washington, and Operational Groups in the field in planning, training, and operating.

FOR OG, WASHINGTON

1. *PROCUREMENT OF PERSONNEL*

a. Is the request for procurement and training of personnel for OG's approved by proper authority?

b. Does the allotment of officers and enlisted men to OG permit the procurement of the numbers requested by the strategic services officer?

c. Are qualified individuals available in the U.S. Army?

d. Can suitable personnel be procured in the time available?

e. Have detailed requests been submitted to the Personnel Procurement Branch, OSS, for procurement of personnel?

f. Is the OSS area in which the OG's are initially to be received properly staffed and equipped to receive them?

SECRET

g. Have the required numbers of suitable personnel been procured and dispatched to the holding area?

h. Have personnel been procured for Field Service Headquarters?

2. *TRAINING*

a. Have all members of the OG's received basic military training?

b. Are suitable OG instructors available and assigned?

c. Has training schedule been coordinated with Schools and Training Branch?

d. Does the standard curriculum for OG's require addition of specialized training for a particular Group? If so, where is it to be accomplished?

e. Is the training area prepared to receive the OG's?

f. Are there any unqualified or unsuitable individuals who should be dropped from the OG's? Are replacements available?

g. Has training accomplished its objectives?

h. Are abilities properly recognized by assignment of ranks and grades within the groups?

i. Has the strategic services officer been informed of the training given OG's to be assigned to his theater?

j. Has personnel of Field Service Headquarters been given adequate training?

3. *SUPPLY AND EQUIPMENT*

a. Has each member of the OG's complete standard army clothing and equipment and special items of individual equipment prescribed?

b. Has each OG complete equipment as prescribed by its approved T/E?

c. Has each Field Service Headquarters the equipment and supplies prescribed by its T/E?

d. Has each Area Headquarters the equipment and supplies prescribed by its T/E?

e. Is special OSS equipment required for the OG's available in the theater?

(1) If so, have requisitions been received and when will the equipment and supplies be shipped? Has the strategic services officer been given complete information?

f. Is any equipment requisitioned unavailable? When will it be available? Has the strategic services officer been informed?

g. What is the schedule of future shipments of supplies and equipment?

h. Is a Table of Equipment sent to the Port of Embarkation with each OG?

4. *MORALE*

a. What is the state of morale in the OG's during training?

b. Is personal contact maintained with the trainees and are facilities available for handling individual morale cases?

c. Are the trainees conscious of the seriousness and the importance of the work?

d. Are periods for rest, relaxation, and diversion provided?

e. Has the schedule been arranged so that there will be no prolonged periods of idleness?

f. Will the OG's depart for the theater promptly after the training period? If early departure is impossible have further training or useful duties been scheduled?

5. *SECURITY*

a. Has each member of the OG's received a security check while he is at the holding area and prior to his specialized OG training?

b. Has the trainee evidenced a sufficient appreciation of security in training?

c. Has each member of the OG's received a security check for overseas service?

6. *TRANSPORTATION*

a. Have all arrangements been completed to transport the OG's to the theater promptly after training is completed?

    (1) Theater commander's approval?

    (2) T/O's and T/E's complete?

    (3) Inoculations and physical examinations completed?

b. Has an OG roster been sent to the strategic services officer?

c. Has the strategic services officer been informed when additional personnel requested will be transported?

7. *REPORTS*

a. Are reports on OG operations received from the field?

b. Do reports indicate that the operations of OG's conform to approved Strategic Services over-all and special programs?

c. Are the reports from the field complete and in the prescribed form?

SECRET

CHECK LIST

FOR OG's, THEATER

1. *PLANNING*

<u>a</u>. Do the projected operations conform to approved Strategic Services over-all and special programs?

<u>b</u>. Has the operational plan been approved and co-ordinated by proper authority?

<u>c</u>. Is all available intelligence considered and plans kept up to date?

<u>d</u>. Has a system of supply been determined?

2. *PERSONNEL*

<u>a</u>. Is Field Service Headquarters present and organized to administer control over OG's?

<u>b</u>. Are the OG's up to strength? If not, can the required additional personnel be procured in the theater?

<u>c</u>. Is the organization of OG's complete and in conformity with the T/O?

<u>d</u>. Have the personnel of the OG's been inspected individually to determine their morale and physical fitness?

3. *TRAINING*

<u>a</u>. Have the OG's received all specialized training required for the tasks assigned?

<u>b</u>. Has the training of specialists been adequate to enable them to perform their individual duties?

<u>c</u>. Has the training of the personnel of Field Service Headquarters and Area Headquarters prepared these organizations properly to perform their functions?

4. *SUPPLY AND EQUIPMENT*

a. Are the OG's, Field Service Headquarters, and Area Headquarters fully equipped in conformity with the T/E's?

b. Is the required special OSS equipment available? If not, have requisitions been submitted? When will it be received?

c. Is the individual equipment of the OG's complete and in order?

d. Has a detailed supply plan been made for each task?

5. *REPORTS*

a. Have detailed reports, within the limits of security control, been sent to OSS, Washington?

www.ingramcontent.com/pod-product-compliance
Lightning Source LLC
Chambersburg PA
CBHW050251230526
45470CB00005B/2219